Industrial Agriculture

I. Industrial Agriculture Effects on Humans and the Environment
II. Converting a NJ Crop Farm to Renewable Energy

Nagwa Awad

ELIVA PRESS

ELIVA PRESS

Nagwa Awad

Agriculture is the single most important practice for human survival. But farms have progressed from the small family owned farms to large Industrial Farms (IF) with machinery. This trend has led to decline in human health, environmental degradation and a surge in poverty.

This book presents a solution. By incorporating renewable energy, farms can reduce cost. Human health and wealth will see a steady improvement and the environment will be on a steady road to recovery.

Published: Eliva Press SRL
Address: MD-2060, bd.Cuza-Voda, 1/4, of. 21 Chişinău, Republica
Moldova
Email: info@elivapress.com
Website: www.elivapress.com

ISBN: 978-1-63648-021-3

I. Industrial Agriculture Effects on Humans & the Environment

II. Converting a NJ Crop Farm to Renewable Energy

Nagwa Awad

Table of Contents

I. Industrial Agriculture Effects on Humans & the Environment 3

Executive Summary 3

Industrial Agriculture Economic Evaluation 4

Industrial Agriculture and Poverty 8

Industrial Agriculture and Health Impact 11

Industrial Agriculture and Water 12

Industrial Agriculture and Fossil Fuel 14

Conclusion & Recommendations 17

References Part I 19

II. Converting a NJ Crop Farm to Renewable Energy 21

Executive Summary: 21

Project: 23

    Farm Profile: 23

    (PART 1) SOLAR: 25

        Estimated Annual Operating cost: 26

    (PART 2) WIND: 27

        Capital Cost Estimates: 27

        Estimated Annual Operating Cost: 28

        Estimated Production Cost: 28

        Projected Selling Price: 28

Barriers & Hindrances: 28

Potential Strategies for Selling Energy: 29

Health and Environmental Benefits Estimates: 30

        Greenhouse Gas Emissions Reduction: 30

        DALY: 30

        Species Damage Reduction: 30

        Scarcity Damages: 30

Reference Part II 31

I. Industrial Agriculture Effects on Humans & the Environment

Executive Summary

Humans began manipulating and domesticating plants in the 1800's (National Agricultural

Library, 2010). Today, farming is an industry. This trend in agriculture led too much use of

fossil fuel, too little access to water and an imbalance of food security. Other problems resulting

from such policies are hunger and poverty. There doesn't seem to be a balance to access to

nutritious food. With poverty and hunger being the first two goals of the United Nations

Sustainable Development Goals (SDG) (Sustianable Development Goals, 2018), and with 86%

of the world's energy source coming from fossil fuel (World Energy Council, 2016), examining

farms and the farming industry here in the United States to determine where the shortfalls and

gaps exist that lead to an imbalance in the nexus of food, water and energy can help. We have

turned food into non-food essentials while children go to bed hungry. This paper will first

present data about the state of large and small farms and then conclude by addressing ways in

which we can use farming to provide more food to the poor, conserve water and reduce energy

emissions.

Industrial Agriculture Economic Evaluation

Agriculture worldwide has a 3.8% share of the Gross Domestic Product (GDP) (The World Bank, 2015). It is the single largest employer in the world with 40% employed in agriculture. (United Nations Sustainable Development Goals, 2017). One third to one half of the world population are income and wealth poor farmers who rely on farming and herding for their livelihood while 3 to 4% of the world's agriculture population are industrialized and technology advanced farms that represent roughly 66% of the value of world agricultural output between 1995-1997 (de Vries, 2013). About 500 million rain-fed farms provide 80% of food in developing countries (United Nations Sustainable Development Goals, 2017). With more than 50,000 edible plants, only wheat, rice and corn directly or indirectly provide 80-90% of all the calories that humans consume (Bourne, Jr, 2015). Since the 1900's 75% of crop diversity has been lost (United Nations Sustainable Development Goals, 2017).

Currently, the world is producing less of such crops as wheat, rice and nuts and a fractional increase of between 0.2% and 1% of corn, oats and barley (World Agricultural Production, 2018). This cannot sustain a world growing at a rate of 1.1% (United Nations, 2015). The only crop being grown at a rate keeping pace with population growth is soybean which America currently dominates (World Agricultural Production, 2018), but given current political climates, this might not be the case next year.

Agriculture and related industries, in America, such as food processing and manufacturing contribute $992 billion to the Gross Domestic Product (GDP) in 2015, about 5.5% (USDA, 2017). Small farms, those who have gross cash farm income (GCFI) of less than $350,000 account for about 90% of U.S. farms and contribute about one fourth of the value of production while large farms with a GCFI of $1 million or more account for less than 3% of U.S. farms and contribute 45% of the value of production (United States Agricultural Department, 2017). Small farms have to rely on off-farm income and micro-farms (those producing less than $10,000 in farm income) actually see a negative rate of investment (ROI) where their off-farm income is greater than their total income (United States Agricultural Department, 2017). About 82% of farm income comes from off-farm employment (Bunge & Newman, 2018). Figure 1 shows the concentration of those working off the farm by age. Small farms and farm land is vanishing. A decline in the number of farms (2.2 million in 2007 vs 2.1 million in 2012) and farm land (922million acres in 2007 vs 914million acres in 2012) is indicative of a trend towards less open spaces (United States Department of Agriculture, 2012). Yet farm sizes are increasing from 418 average acres per farm in 2007, to 434 average acres per farm in 2012 (United States Department of Agriculture, 2012). Farm land value is also increasing from an average farm land value of $790,000 in 2007 to and average land value of $1.1 million in 2012 (United States Department of Agriculture, 2012). Industrialization and commercialization of farms have pushed land value beyond the reach of farmers and towards large

Farms with Principal Operator whose Primary Occupation Is Not Farming, by County, 2012

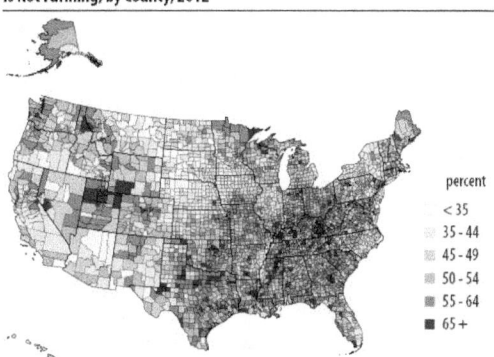

percent
< 35
35 - 44
45 - 49
50 - 54
55 - 64
65 +

Source: USDA NASS, 2012 Census of Agriculture.

Figure 1: Percent of farmers supplementing farm income.

corporations as the average per acre value increased from $1900 in 2007 to $2500 in 2012 (United States Department of Agriculture, 2012). Figure2 shows increase in farm sizes.

As of 2012, Corn and soybean accounted for 50% of total farm output (Census of Agriculture, 2014). There was as an increase in 2012 of corn for silage and soybean for export totaling 40%, while food produced for consumption, such as rice, barley and vegetables fell by

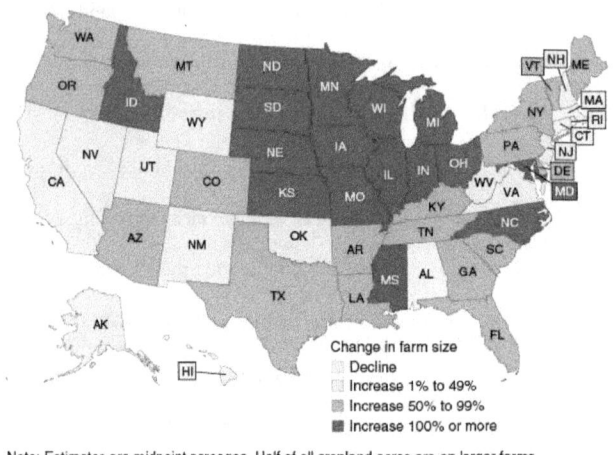

55% (Census of Agriculture, 2014). Corn, wheat and soybean account for 90% of American crops (World Agricultural Production, 2018). The US has 96 million acres of corn and is the leading corn producer (Brown, 2012) with 32% of the world's output of corn, 50% of the

Note: Estimates are midpoint acreages. Half of all cropland acres are on larger farms, and half are on smaller.
Source: USDA, Economic Research Service calculations from unpublished Census of Agriculture records, 1982 and 2007.

Figure 2: Percentage increase in farm sizes across the nation

world's output of soybean and places third in global wheat output (National Agricultural Library, 2010). However, in 2012 corn harvest was hit with an intense heat wave that affected global food prices (Brown, 2012). The world grain reserves fell by one third while world grain prices more than doubled between 2007 and 2008 (Brown, 2012). Despite America having stock reserve, in the past, and idled farms, currently the nation does not have stock reserve and very few land for idling, which contributes to spikes in food prices even though corn yield has not yet plateaued (Brown, 2012). In 2012, for the second year in a row, the global food price index rose by 1.1% (United Nations Sustainable Development Goals, 2017). Currently, however, the US is

the largest importer of vegetables and it is expected that the country's imports will be greater

than its exports by 2050 (CLIMATE CHANGE GLOBAL FOOD SECURITY amd the AND

U.S. FOOD SYSTEM, 2015). To avoid a recurrence of the 1980 farm financial collapse

(National Agricultural Library, 2010), farms are heavily subsidized by the government. With the

looming trade war that would affect America's soybean export, and given the heavily

government subsidized agricultural industry, the nation's farms are doomed to repeat the

agricultural collapse that ended in 1981 that had shaped the farming industry as a government

dependent, industrial sized farms that predominantly produce cattle feed grain and biofuel

instead of human feed grain (National Agricultural Library, 2010). Figure 3 shows farm output

by state.

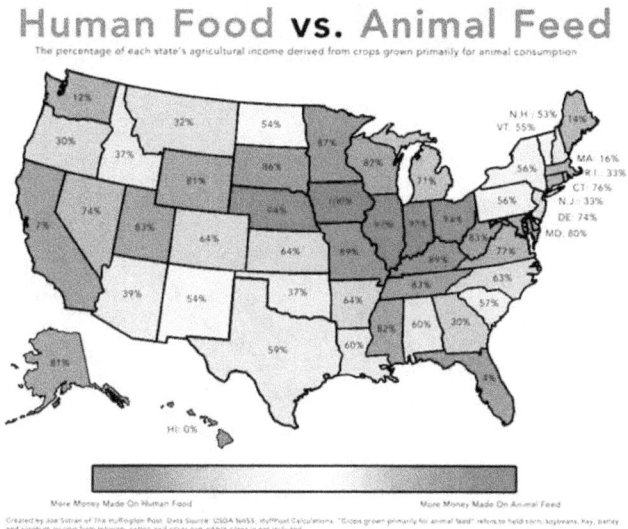

Figure 3: Farms producing food for humans and farms producing food for feed by state

Industrial Agriculture and Poverty

A byproduct of current economic agricultural trends is the rise in poverty and hunger.

Poverty is defined as the inability to afford basic needs such as food, clothing and shelter

(UNESCO, 2017). Income Poverty is *"when a family's income fails to meet a federally*

established threshold" (UNESCO, 2017, p. 1 para. 1) such as the case for small farms in

America. Interestingly, the US ranks 7th in citizens living below the poverty line with poverty

rate of 29.4% (Graph 1) (Bruckauf, Chzhen, Cuesta, Richardson, & Toczydlowska, 2017). As

previously stated, one third to one half of the world's populations are income and wealth poor

farmer. Yet in America, farming accounts for only 2.8% of the labor market (National

Agricultural Library, 2010), while farm sizes are getting larger; leading one to conclude that

industrial farming has caused the drop in farming labor. For farmers, crop provides subsistence

and income and agriculture drives rural development (Peralta & Hunt, 2003). Farming as a

means of support has led to poverty among farmer.

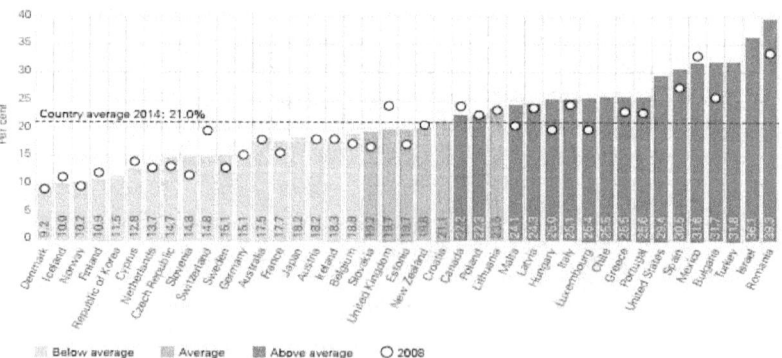

Graph 1: is a list of rich countries and the percentage of children under 18 that fall below the country's poverty lines.
(Bruckauf, Chzhen, Cuesta, Richardson, & Toczydlowska, 2017).

The U.S. Census Bureau reports 12.7% poverty in the U.S. with a poverty threshold of $24,339 for a family of 2 adults and 2 children for 2016 (U.S. Census Bureau, 2016). Notably this figure is much lower than UNICEF's figure of 29.4%. As previously discussed, small farms account for 90% of farms in America, but only produce less than 25% of output. Small farms rely on off-farm income as they attempt to compete with large industrialized farms. This also adds stress to the labor market as farmers compete for jobs in all sectors. Only about half of the farms in America are owned (Census of Agriculture, 2014). Net income from farms peaked at $50,000 a year in 2012, but has since been declining to about $38,000 a year (Bunge & Newman, 2018), which is close to the $25,000 poverty line for a family of four. This does not take into account the projected increase in fertilizer and seed cost (Bunge & Newman, 2018) that will further reduce the net income from farms. According to studies, people living in poverty spend 50-70% of their income on food (Brown, 2012). When prices rise, hunger is the end result with currently 20% of Americans struggling to provide food for their families (Hanlon, Madel, Olson-Sawyer, Rabin, & Rose, 2013). Just like the poor and due to the depletion of food reserves, the world is living one year to the next (Brown, 2012). Compounding the situation is the fact that 35% of harvested grains are used for cattle feed and not fit for human consumption which further causes severe food price inflation (Brown, 2012). It's well known that poverty inhibits the ability to produce income which increases poverty (Peralta & Hunt, 2003). It is also a fact that the poor are more likely to be harmed by poor quality of air, food and water (Peralta & Hunt, 2003). Overall, an estimated 815 million people are poor and undernourished with that number projected to increase to two billion by 2050 (United Nations Sustainable Development Goals, 2017). Figure 4 shows poverty throughout America.

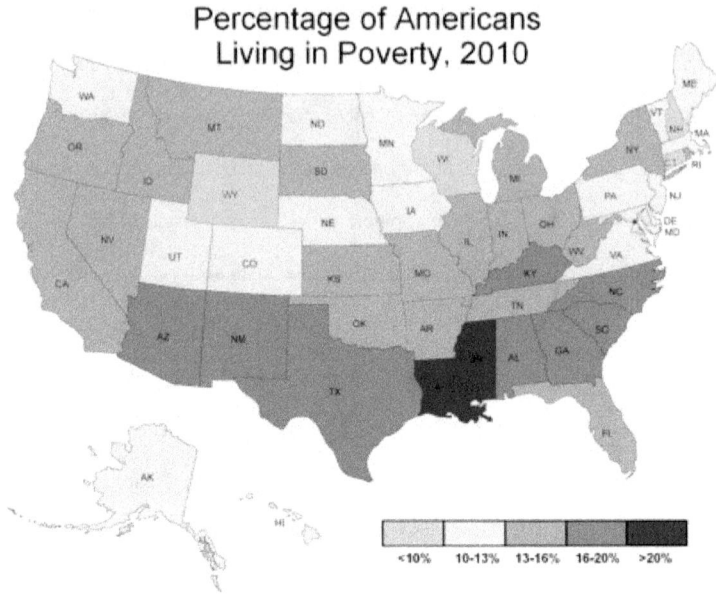

Figure 4: Percentage of American living in poverty by state

Children are affected by poverty more poignantly than adults. Children experience poverty as an environment that is damaging to their mental, physical, emotional and spiritual development (UNICEF, 2005). Poverty in children leads to going without a nutritious meal that would contribute to a healthy growth, or being subjected to hazardous labor that prevents access to education among other things (UNICEF, 2005). Thus, preventing stimulation or emotional support that can inhibit potential (UNICEF, 2005). About an average of 20% of children in rich countries live in Relative Poverty and more than 10% either go hungry or don't get a nutritious meal (Edmond, 2017). An estimated 25% of children's growth is stunted by poverty and about half of deaths in children under the age of five are attributed to malnutrition (United Nations

Sustainable Development Goals, 2017). Children born into poverty are twice as likely to die by age five (United Nations Sustainable Development Goals, 2017).

Industrial Agriculture and Health Impact

Food technology innovations introduced the world to Genetically Modified Organisms in 1996 (GMOs) which are defined by the World Health Organization (WHO) as biotechnology that alters food in a way that does not occur naturally (World Health Organization, 2014). One of the purposes was to increase crop production by introducing disease resistant plants (World Health Organization, 2014, p. 1 para. 4). Such undesirable control by a few chemical companies led to their dominance of agricultural development and an unsustainable agricultural economy (World Health Organization, 2014). Other concerns pertain to human health issues. WHO reported three areas of particular concerns: allergenicity or the transfer of genes from allergic organism to non-allergic organism; gene transfer from GM's to the human body might impact human health, and outcrossing or the migration of GM plants into conventional crops. (World Health Organization, 2014). GMO's were also intended to produce insect and disease immune crops, but looking at statistical output indicates a negative output. These agricultural economics and human health concern render these trends in the global agro-food systems unsustainable, and with food production decreasing while the world population is increasing, this can lead to collapse in several societies globally. In 1996, biotechnological advances introduced genetically engineered (GE) seeds for crops that led to a use in most crops of 90% (USDA, 2016). GE crops are defined by the Economic Research Services (ESR) of the United States Agricultural Department (USDA) as "plants with genetic material that has been altered" (McFadden, 2017). GE crops are herbicide-tolerant, insect-resistant crops that can lead to a reduction of farmers' agricultural expenses (McFadden, 2017). While GE crops seem to save the American farmer resources, there is a down side. A study found that GE's or Genetically Modified (GM) foods toxically stress human organs and can lead to *"toxic effects such as hepatic, pancreatic, renal, or reproductive effects and may alter the hematological, biochemical, and immunologic parameters" (Dona & Arvanitoyannis, 2010, p. 1 para. 1).*

Industrial Agriculture and Water

Water is a vital aspect of the agricultural system dynamic which is essential for humans, crops and animals and for many industrial processes (Water Security: The Water-Food-Energy Nexus, 2011). According to International Atomic Energy Agency (IAEA), 70% of the world is covered by water but only 2.5% of it is fresh, only 1% of fresh water is available for human and ecosystem use and only about 0.3% is easily accessible (Internationa Atomic Energy Agency, 2011). About 70% of fresh water is used for agriculture and another 8% is used for domestic consumption (Figure 5); it is estimated that about 85% of rainwater does not reach crops (Internationa Atomic Energy Agency, 2011). With rising temperatures, globally, we are experiencing soil erosion and water shortages (Brown, 2012). Additionally, over 80% of waste water discharge into surface water is unfiltered (United Nations Sustainable Development Goals, 2017).

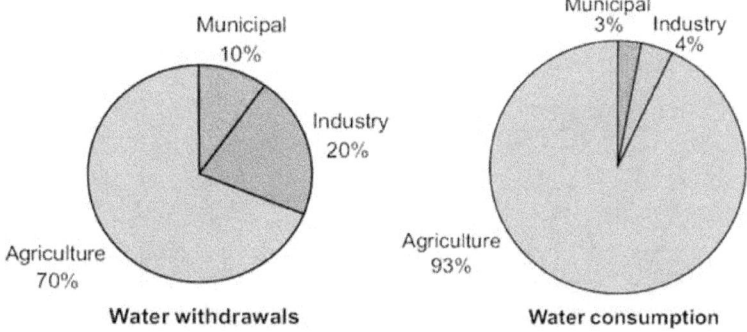

Figure 5: Water consumption in America by sector

In America it's estimated that agriculture accounts for 80% of the overall nation's water consumption and about 90% in the Western States (Schaible & Aillery, 2017). Of course with higher percentage than the worldwide of 70%, efforts are being made by the USDA into improving farm water management, dry-year water banks and adopting irrigation technologies that can stem the impact of irrigated production (Schaible & Aillery, 2017). Such statistics lead towards an unsustainable model of farming and water usage. Natural disasters, such as the dust bowl, can further stress water resources. The Dust Bowl was a series of dust storms following extreme drought in the Midwest and Southern Plains between 1931 and 1939 (The Dust Bowl). Stress to the agricultural system didn't end until the rains came (The Dust Bowl). Industrial agriculture has led to excessive irrigation and water salinity (Brown, 2012). America currently is one of the leading nations in over pumping of aquifers (Brown, 2012).

Industrial Agriculture and Fossil Fuel

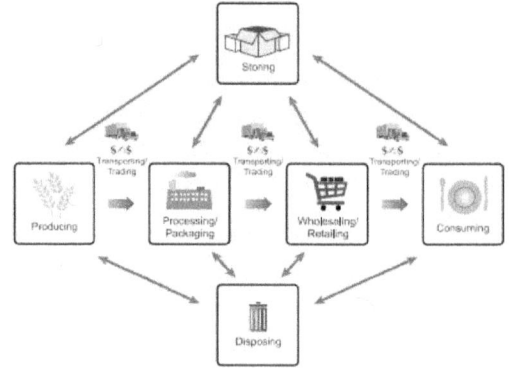

Figure 6 shows the need for energy in all steps of industrial agriculture. As previously stated, about two-thirds of the world's agricultural output is by about 3-4% of farms. Such farms

Figure 6: Flow diagram of agriculture process

are at the center of the world in their advancement in industrial economies, but their relationship with natural ecosystems has almost vanished (Morris, 2015). One of the key characteristics of Industrial agriculture is intensive use of external inputs such as fossil fuel (Morris, 2015). Non-renewable minerals and fossil fuel are two vital resources to industrialization (de Vries, 2013). Currently 86% of energy source worldwide comes from fossil fuel (World Energy Organization, 2016). Less than 3% of American farms produce almost 50% output. Coupled with the shrinkage of the farm sizes, previously discussed, leads to the conclusion that small labor centered farms are being replaced with large industrialized, centralized farms. With industrialization comes fossil fuel causing air pollution emitted from machinery causing health risks and contributing to global warming (Grace Communications Foundation, 2016). Currently the U.S. generates 65% of electricity using fossil fuel and only 15% come from solar energy (U.S. Energy Information Administration, 2017). Fossil fuel as a source is high in energy and crucial to humanity. But due to the growing trend of industrial farming, more Green House

Gasses (GHG) are emitted into the air leading to rising temperatures which cause rising food

prices (Brown, 2012) in addition to health problems. Such policies lead to climate change which

would affect food security (CLIMATE CHANGE GLOBAL FOOD SECURITY amd the AND

U.S. FOOD SYSTEM, 2015).

Table 1		
Environmental Systems Statistics Worldwide		
Agriculture	Water	Fossil Fuel
3.8% of GDP from agriculture	70% of the world is covered by water	Industrialized farms rely on fossil fuel
25% live in agricultural villages	2.5% of the water is fresh; 1% is available, 0.3% accessible	Fossil fuel is finite and is depleted upon exploitation
30% depend on farming & herding for livelihood	1 billion people don't have access to safe drinking water	Supplies 86% of world energy
50% of farmers are poor	70% of fresh water used for agriculture; 8% for domestic consumption	Environmental sinks are finite
3-4% of industrialized farms produce 66% of output	85% of rain water does not reach crops.	

Table 1: Summarizes key points that lead to agriculture, water and fossil fuel systems being unsustainable.

Table 2		
Environmental Systems Statistics in America		
Agriculture	Water	Fossil Fuel
$992B or 5.5% of GDP from agriculture	80% used for irrigation nation wide	Over use of machinery leads to air pollution
90% farms with GCFI < $350K >3% GCFI >= $1M produce 45% of output GCFI =< $10K rely on off-farm income with a negative ROI	90% used for irrigation in Western States.	65% of electricity source
Farms 2007: 2.2M 2012: 2.1M Acres 2007:922M 2012:914M		
90% corn crop is GE seeds		

Table 2: Summarizes key points that lead to agriculture, water and fossil fuel systems being unsustainable.

.

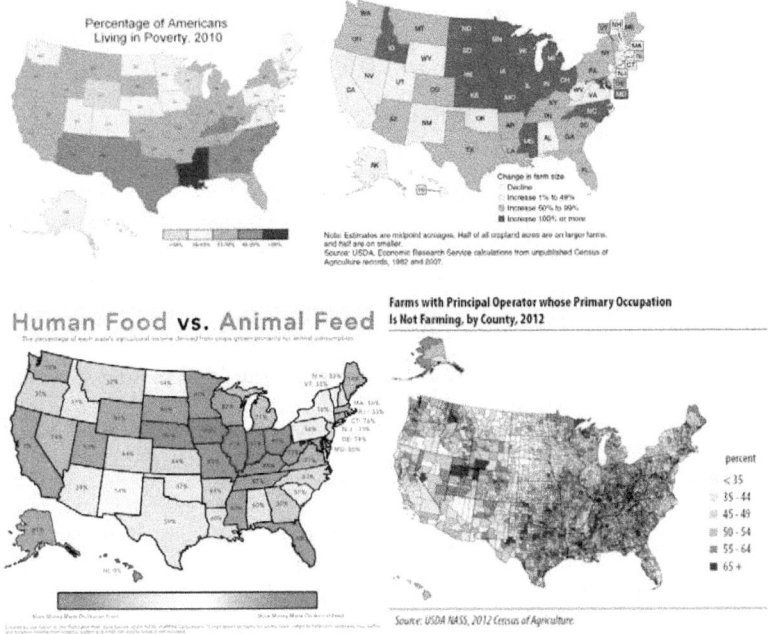

Figure 7: Side by side comparison of poverty, farm size, farm output and farm supplemental income

Conclusion & Recommendations

There is no doubt that there has been a rapid increase of industrial farming since 1960. It is evident that industrial agriculture has a direct impact on rural poverty. Looking at the maps of America side by side (Figure 7) shows a direct correlation between poverty, farm size and farm type (crop for feed or crop for humans). The economic impact industrial farming has on small rural farms has been devastating. But with the US imposing more stringent auto fuel efficiency, this can lead to more use for electric cars on the farm at a cost of $0.80 a gallon (Brown, 2012). Such measures can lead to economic benefits such as more food for humans being harvested and lower food prices (Brown, 2012). Other examples of government stepping in, is that oversight agencies have been established to manage water use in states such as California (Hanlon, Madel, Olson-Sawyer, Rabin, & Rose, 2013). These policies would ultimately lead to small farmers experiencing better economic prosperity. There are positive aspects of industrial farming and the use of GMO. With new technologies, industrial farming can mass produce food fortified with iron and zinc (Peralta & Hunt, 2003). Such is the case with golden rice which is fortified with beta-carotene (Bourne, Jr, 2015). These measures can reduce health issues with malnutrition and reduce hunger. Other positive steps would be to address issues with food waste. Currently, 150-300 billion pounds of food is wasted per year costing $165 billion annually (Hanlon, Madel, Olson-Sawyer, Rabin, & Rose, 2013). Food waste makes up 40% of total food output at the processing and consumer level in industrial countries (United Nations Sustainable Development Goals, 2017). But if we wasted 5% less we can feed 4 million more Americans per year, and wasting 15% less food would feed 25 million more people annually (Hanlon, Madel, Olson-Sawyer, Rabin, & Rose, 2013). Other options would be to encourage women farmers which would reduce hunger by half (United Nations Sustainable Development Goals, 2017). When food insecurity diminishes, global economic productivity increases by 2-3% annually (CLIMATE CHANGE GLOBAL FOOD SECURITY amd the AND U.S. FOOD SYSTEM, 2015).

Ending industrial agriculture requires studies beyond the scope of this paper, but we looked at ways to work with the existing system. We discussed ways to improve small farmers' economic position by managing water rights, reducing poverty and hunger. We presented ways to use industrial agriculture to improve the health impact. We identified measures the US is requiring to reduce emission output. All these policies, while seeming minor, would have a large

impact on the farming industry, food, water resources and energy output.

References Part I

Bourne, Jr, J. (2015). *The End of Plenty: The Race to Feed a Crowded World.* New York: W.W. Norton & Company.

Brown, L. R. (2012). *Full Planet, Empty Plates.* New York: W.W. Norton & Co.

Bruckauf, Z., Chzhen, Y., Cuesta, J., Richardson, D., & Toczydlowska, E. (2017). Florence, Italy: UNICEF Office of Research Innocenti.

Bunge, J., & Newman, J. (2018, February 25). *To Stay on the Land, American Farmers Add Extra Jobs.* Retrieved from The Wall Street Journal: https://www.wsj.com/articles/to-stay-on-the-land-american-farmers-add-extra-jobs-1519582071

Census of Agriculture. (2014, September). Retrieved from U.S. Department of Agrigulture: https://www.agcensus.usda.gov/Publications/2012/Online_Resources/Highlights/Farms_a nd_Farmland/Highlights_Farms_and_Farmland.pdf

CLIMATE CHANGE GLOBAL FOOD SECURITY amd the AND U.S. FOOD SYSTEM. (2015, December). Retrieved from USDA.ORG: https://www.usda.gov/oce/climate_change/FoodSecurity2015Assessment/FullAssessment .pdf

de Vries, B. J. (2013). *Sustainable Science.* New York: Cambridge University Press.

Dona, A., & Arvanitoyannis, I. S. (2010, January). *National Center for Biotechnology Information.* Retrieved from U.S. National Library of Medicine: https://www.ncbi.nlm.nih.gov/pubmed/18989835

Edmond, C. (2017, June 28). *WEF.* Retrieved from The Rich Countries have High Childhood Poverty : https://www.weforum.org/agenda/2017/06/these-rich-countries-have-high-levels-of-child-poverty/

Grace Communications Foundation. (2016). Retrieved from Sustainable Table: http://www.sustainabletable.org/265/environment

Hanlon, P., Madel, R., Olson-Sawyer, K., Rabin, K., & Rose, j. (2013, January). *Fodd, Energy, Water: Know the Nexus.* Retrieved from http://www.gracelinks.org/media/pdf/knowthenexus_final_051513.pdf

Internationa Atomic Energy Agency. (2011, September). Retrieved from IAEA: https://www.iaea.org/sites/default/files/publications/magazines/bulletin/bull53-1/53105911720.pdf

McFadden, J. (2017). *USDA.* Retrieved from The United States Department of Agriculture; Economic Research Services: https://www.ers.usda.gov/amber-waves/2017/august/although-small-markets-have-been-expanding-for-ge-crops-with-traits-that-increase-nutrient-content-or-improve-taste/

Morris, I. (2015). *Foragers, Farmers and Fossil Fuel.* Princeton: Princeton University Press.

National Agricultural Library. (2010). Retrieved from United States Department of Agriculture: https://www.nal.usda.gov/agricultural-history

Peralta, G. L., & Hunt, J. M. (2003). *A Primer on Helath Impacts of Development Programs.* Asian Development Bank.

Schaible, G., & Aillery, M. (2017, April 28). *The United States Agricultural Department.* Retrieved from Economic Reseach Services: https://www.ers.usda.gov/topics/farm-practices-management/irrigation-water-use.aspx

Sustianable Development Goals. (2018). Retrieved from United Nations SDG: https://www.un.org/sustainabledevelopment/sustainable-development-goals/

The Dust Bowl. (n.d.). Retrieved from PBS: https://www.pbs.org/wgbh/americanexperience/features/dust-bowl-surviving-dust-bowl/

The World Bank. (2015). *Agriculture, value (% of GDP).* Retrieved from https://data.worldbank.org/indicator/NV.AGR.TOTL.ZS

U.S. Census Bureau. (2016). Retrieved from The United States Census Bureau: https://www.census.gov/topics/income-poverty/poverty.html

U.S. Energy Information Administration. (2017, May 10). Retrieved from https://www.eia.gov/energyexplained/index.cfm?page=electricity_in_the_united_states#tab2

UNESCO. (2017). Retrieved from United Nations Educational, Scientific and Cultural Organization: http://www.unesco.org/new/en/social-and-human-sciences/themes/international-migration/glossary/poverty/

UNICEF. (2005). Retrieved from United Nations Emergency Fund: https://www.unicef.org/sowc05/english/povertyissue.html

United Nations. (2015, September 25). Retrieved from http://www.un.org/sustainabledevelopment/sustainable-development-goals/

United Nations Sustainable Development Goals. (2017). Retrieved from United Nations: http://www.un.org/sustainabledevelopment/sustainable-development-goals/

United States Agricultural Department. (2017, November 29). Retrieved from Economic Research Services: https://www.ers.usda.gov/data-products/ag-and-food-statistics-charting-the-essentials/farming-and-farm-income/

United States Department of Agriculture. (2012). Retrieved from Census of Agriculture: https://www.agcensus.usda.gov/Publications/2012/Full_Report/Volume_1,_Chapter_2_US_State_Level/st99_2_008_008.pdf

USDA. (2016, August 22). Retrieved from the United States Agricultural Department; Economic Research Services: https://www.ers.usda.gov/topics/farm-practices-management/

USDA. (2017, October 18). Retrieved from The United Stated Department of Agriculture; Economic Research Services: https://www.ers.usda.gov/data-products/ag-and-food-statistics-charting-the-essentials/ag-and-food-sectors-and-the-economy/

Water Security: The Water-Food-Energy Nexus. (2011). Washington, D.C.: Island Press.

World Agricultural Production. (2018, July). Retrieved from United States Agricultural Department: https://apps.fas.usda.gov/psdonline/circulars/production.pdf

World Energy Council. (2016). *World Energy Resources*, 1028.

World Energy Organization. (2016). *World Energy Resources*, 1028.

World Health Organization. (2014, May). Retrieved from Food Safety: http://www.who.int/foodsafety/areas_work/food-technology/faq-genetically-modified-food/en/

II. Converting a NJ Crop Farm to Renewable Energy

Executive Summary:

New Jersey (NJ) has an abundance of sun in the summer and an abundance of wind in the winter. This project is about installing a 2.6 MW solar PV system and a 1.4 MW 3K wind turbine system on a crop farm in Middletown Township, Monmouth County, NJ, USA. Sometimes I will be using data for the township, the county, the state or the country. The farm is 47 acres. I will be allotting 5 acres for the solar PV system. The wind turbines will not require much area and can be installed among the crop areas. Electricity expenses for the farm are $12,935 per year. The farm uses 6521 KWh per month at a rate of $0.164 per KWh. Current net annual income for the farm is $13,293 per year from sales and government grants. The electricity mix for the farm is about 50% from natural gas and about 40% from nuclear. The farm owners are initiating the project and to secure funding will be using the property as collateral valued at $986,154 with additional loans and guarantees from the USDA. The owners are able to secure loans at a rate of 2% for a 30 year fixed term.

Methods: I used a manual calculator and the calculators provided by Dr. Sanchez to double check my results. In the case of wind, I included my calculations but decided to use the total energy output from the calculator provided in class since that yielded more output.

Solar: PV panels 150 W/m^2 efficiency will be installed over 17,500 m^2 at a cost or $1.83 per watt yields a total production cost of $7,869,240 (subtracting for a $600,000 REDA grant) and a total annual production of 3,363,874 KWh. Annual mortgage payments plus an estimated 1% of production cost yearly expense equals an annual expense of $430,053. Production cost is 11 cents per KWh which allows the farm to make a profit at a 10% or 20% margin. Potentially, the farm can make $80,740 of profit (that includes money saved from current electric expenses).

Wind: A total of 466 3KW 4 meter wind turbines will be installed among the crops. NJ meets the minimum threshold for year round energy output from wind (3.99 m/s). I calculated the estimated installed capacity for both a utility and residential project, but concentrated on this being predominantly a large scale residential project. Using the calculator from class, and inputting average wind speeds for each month, we arrive at a total annual production of 6,852,959 KWh. At an installation cost of $4.4 per watt and assuming a 3% annual production cost yields total annual expense of $459,844. Production cost is $0.067 KWh. Potentially at a rate of 10% or 20% profit margin, the farm can make $91,830 a year which when compared to the current annual net income of $13,293 makes this investment financially advantageous.

Some barriers for the farm are the lack of access to distribution lines, power lines are exclusively owned by one company that has set prices at which they will purchase electricity, NJ cancelling incentives for renewable energy, the farm losing its low tax liability, large liability insurance expenses and the lack of qualified installers for wind turbines in NJ. Given the scale of the project, the best option for the farm is to sell electricity to townships and coops similar to the programs initiated in New York. The farm can also participate in NJ's cap-and-trade program for added revenue.

Some health and environmental benefits include a reduction in DALY by 16.2 year, and scarcity damages of about $15,988,564.

The farm can benefit from converting to renewable energy because not only will it become self-sustaining, it stands to gain an additional $172,407 in yearly revenue in addition to their current income of $13,293 making this a financially worthwhile investment.

Project:

My project is about a theoretical crop farm in Middletown Township, in Monmouth County, New Jersey. I'm using characteristics of the farm from averages obtained from the United States Department of Agriculture, National Agriculture Statistics Service (USDA, 2017), Rutgers University (Both, Manning, & Rabin, 2014), U.S. Energy Information Administration (EIA, 2019) and Electricity Local (Electricity Local). I will be looking to mete out 5 acres for a 2.6 MW solar system and to install 1.4 MW wind system among the 42 acres without having to lose crop area. The systems I plan to install are solar PV and 3KW wind turbines. Excess electricity will be sold and carbon will be traded in NJ.

Farm Profile:

Size: 47 acres; land value: $986,154 ($20,982 per acre * 47 acres); *Total farm production expenses*: $97,971 (Figure 1)

Electricity Expenses: $12,935 ($0165 * 6521 KWH * 12 months); Price per KWH for commercial: $0.165; price per KWH for residential: $0.164 (Figure 2)

Net Income: $13,293 annually (net cash farm income: $11,630 per year + government payments $1663 per year) (Figure 3). This translates to an annual reduction in net income of $1981 (($13,293/47 acres)*5 acres)

Carbon Emissions (NJ): 111.4 Million Metric Tons (EIA, 2018)

Maps: *For Solar*: I will use the Global Horizontal Global Irradiation 1998-2016 (Solar Maps, 2016) (Figure 4); *For Wind*: I will use the Land based and off-shore U.S. 100-m wind speed map (Wind Map) (Figure 5).

Source for Composition of Electricity Mix: Information source will be gathered from the Energy Efficiency and Renewable Energy website. The information will be specific to New Jersey (they do not have county wide). They allow a download into a spread sheet which I then programmed to show percentages of output (Figure 6) (EIA, 2018). ***Monmouth County Electricity Mix:*** Biomass: 1.25%; Wind: 0.03%; Solar: 4.19%; Hydroelectric (conv): 0.05%; Geothermal: 0%; Natural Gas: 50.54%; Nuclear: 41.67%; Coal: 1.55% (EERE, 2018)

Electric Grid: To find the closest power line for possible future sale of electricity, I used the interactive map from the department of homeland security and discovered there are only two substations nearby on the electric grid for NJ (DHS). Substation "Atlantic" and substation "Freneau" would make selling electricity to the grid feasible (see maps). I downloaded the spread sheet and removed any "inactive" or "out of service" entries and anything with a "-99999" voltage (Figure 7),

Land Rights: This project will be initiated by the owner of the land. For mortgage purposes, the owner will use the farm (valued at $986,154) as collateral.

Financing: I wanted to explore grants as a way to minimize the financial capital impact on the farm owner. NJ has suspended renewable energy agriculture grants temporarily (NJ BPU), but there are a few options through the federal government.

Grants:

1) The Rural Energy for America Program (REAP) through the USDA rural development program (RD USDA, 2018) allows the farm to qualify grants of between $1,500 - $500,000 as long as the combination of the two grants exceeds $1,500 and does NOT exceed 25% of the total cost of the project.

2) The Energy Audit and Renewable Energy Development Assistance (EA/REDA) grants (EA REDA, 2019) through the USDA allows the farm to tap into up to $100,000 in funds for assessment and logistical expenses. The farm must show that at least 50% of net income is a result of agricultural production, which is the case.

Green Loans:

Several options in the form of loan guarantees and direct loans are available through: 1) REAP the federal government offers loan guarantees for up to $25 million for renewable energy systems on farms. 2) The Farming Services Agency offers direct lending for family owned farms for new ventures (FSA). Currently 1.625% to 2.000% (Rural Development Lending, 2018); I will be using the 2% rate.

(PART 1) SOLAR:

Capital Cost Estimates:

1) The available land for PV is 5 acres * (4,047 m^2/acre) = 20,235 m^2 * 85% of usable land for installation (allotting 15% for service roads and other structural requirements) = 17,200 m^2; so we will estimate for 17,500 m^2 of available land.

2) Estimated Installed capacity: assuming a 150 W/m^2 pane;

150 W/m^2 * 17,500 m^2 = 2,625,000 W or 2.62 MW ➔ we'll round down to 2.6MW

3) As of 2019; the cost to commercially install solar panels in the U.S. is $1.83 per W or $1830 per KW (nrel, 2018).

The total cost to install a PV system that will generate 2.6 MW over 5 acres is ($1830 * 2,600 KW) = $4,758,000 * $1.78 per Wdc (because the project is in USA) = **$8,469,240**

The Net cost to the farm (subtracting grants) is ($8,469,240 - $600,000 (REDA)) nets: **$7,869,240**

4) Annual Mortgage Payment: $351,361 ($7,869,240 capital cost * 0.0446 mortgage rate) (Figure 8)

Estimated Annual Operating cost: $78,869 (estimating 1% of production cost 0.01 * 7,869,240)

Estimated Total Annual Production: 3,636,874 KWh/year (Figure 9), assuming an 86.1% de-rate factor using Enphase Microinverter Systems (Technical Brief, 2014) (Figure 10).

Production Cost: 11 cents per KW/h (annual mortgage amount $351,361 + annual operating cost $78,692) / annual production 3,636,874 KW/h = $0.118 cents per KW/h).

Projected Selling Price Scenarios:

At a 10% profit (12.9 cents per KWh): ($0.118 * 10%) * 3,636,874 = $42,915 net before tax profit ($33,903 after 21% tax ($42,915 – ($42,915*21%)).

At a 20% profit (14.3 cents per KWh): ($0.118 * 20%) * 3,636,874 = $85,830 net before tax profit ($67,805 after 21% tax ($85,830 – ($85,830*21%)).

Adding the cost of saving on electric utilities would increase our net profit by $12,935 annually to **$46,838** & **$80,740**

(PART 2) WIND:

Because the wind turbines have a very small foot print, they will be installed in the same area as the cropped land without farming disruptions.

Capital Cost Estimates:

1) The farm uses about 6500 KWh per month, but because we hope to sell the excess electricity and because we have enough acreage on the farm, we will base our estimates on 1,000,000 KWh of energy output per month or 1,000 MWh.

2) Average wind speed in Middletown, NJ is 15.66 mph (Figure 11). Lowest wind-speed is 8.9 mph. Converting 8.9 mph to m/s we get: 3.99 m/s which is above the 3 m/s minimum threshold for year round energy output.

   (1 mile = 1609 meters ➔ 8.9 mph*1609 = 14,320 m/h / 3600 s/h ➔ 3.99 m/s)

3) Estimated Installed Capacity: (Mainly a residential project. output may be sold to coop)

   Basing our calculations on a desired output of 1,000,000 KWh per month gives us 1,000,000 KWh/720 hrs. = 1388 KW per day which we will round to 1400 KW (we would need 466 3KW turbines). Measuring at 4 meters, there is ample space. Wind power per month = 0.85 derate factor * 0.295 * 4 m (rotor diameter) * 1.25 kg/m^3 * 15.66 wind speed^3

   Total Energy Production: 6,852,959 KWh per year using the wind calculator (Figure 12))

4) Installed Cost: An individual-type project in the northeast costs $4,400 per KW (Figure 13) (Barbose, et al., 2017). Total installed cost is (1,400 KW * $4,400) = **$6,160,000**

5) Annual Mortgage Payment: The farm does not qualify for additional grants. The farm still qualifies for a low 2% loan guarantee through REDA for 30 fixed rates; the annual capital cost for the mortgage is $275,044 ($6,160,000 * 0.0446) (Figure 8).

Estimated Annual Operating Cost: Will be $459,844. Assuming a rate of 3% of production cost (0.03 * $6,160,000 = $184,800) + $275,044 annual mortgage.

Estimated Total Annual Electricity Output: Would be 6,852,959 KWh per year based on wind speeds obtained for Middletown, NJ (Figure 12)

Estimated Production Cost: Annually is 6.7 cents per KWh (production cost $459,844/6,852,959 KWh = $0.067 or 6.7 cents per KWh)

Projected Selling Price:

At a 10% profit (7.4 cents per KWh): ($0.067 * 10%) * 6,852,959 = $45,915 net before tax profit ($36,272 after 21% tax ($45,915 – ($45,915*21%)).

At a 20% profit (8 cents per KWh): ($0.067 * 10%) * 6,852,959 = $91,830 net before tax profit ($72,546 after 21% tax ($91,830 – ($91,830*21%)).

The cost saving of $12,935 annual in electric utility was already allotted in the solar section.

Barriers & Hindrances:

1. Most distribution lines for the state are on the west side near the Pennsylvania border, but Monmouth County is located on the east side of the state near the Atlantic Ocean and far from most distribution lines. There are only two 220 & 230 volt lines that seem to

crisscross near the farm that may present easy access (Figure 7) Both lines are owned by Jersey Central Power & Light which sets electricity purchase price at 9.4 cents per KWh (Figure 14). It is not financially feasible to submit the solar system on the farm, but the wind system at a production cost of 6.7 cent per KWh would be a profitable venture.

2. New Jersey will cease renewable energy incentives once it reaches a 5.1% electricity source from solar (NJ BPU), currently the rate is 4.19%. In order to overcome this barrier it's best to initiate this project soon to ensure being grand-fathered in.

3. If the farm is not careful to make sure they receive at least 50% of their sales from selling farm production, they can potentially lose their farm status and owe property taxes on 47 acres of land. This can potentially be a loss of $2,392 * 47 acres = $112,424 annually (Figure 15). The farm must register as a solar/wind farm in addition to a crop farm and sell electricity to avoid such a large tax liability.

4. Wind energy is relatively new in New Jersey. To connect to the grid, utility companies require the owner to maintain a $1 M insurance policy which is excessive and can significantly increase annual production cost (USDEP). There is also a lack of properly certified wind turbine installers in the state (USDEP).

Potential Strategies for Selling Energy:

There are two options for selling electricity in Monmouth County and an option for CO2 trade:

1) Community shared electricity or "Community Solar" (Marsh, 2019), also known as "roofless solar" has been initiated with much success in New York using a 2.7 MW solar array (Meehan, 2018) through a Public Purchase Power Agreement. Participant like this

idea because they can pay less for electricity and feel like they are being environmentally responsible without the capital investment of installing solar panels on their rooftops. Currently, the farm produces 2.6 MW of energy which would cover about 350 households (Meehan, 2018).

2) Selling electricity to the Monmouth-Ocean Area Energy Cooperative. Given the low price per KWh for energy produced from the wind turbines on the farm, it would be cost effective for the coop to purchase their energy directly from the farm. Therefore effort should be made to enter the bidding.

3) NJ ranks 16th for CO2 emission (EIA, 2016) (Figure20). Replacing energy source on the farm to renewables reduces carbon emission by at least 1174 Metric Tons of CO2seq/year (Figure 16). NJ participates in the cap-and-trade carbon program (Aulisi, 2005). The farm can participate in the emissions trading program for carbon credits.

Health and Environmental Benefits Estimates: (Figures 17-19) the reductions below are assuming that only one system is implemented.

Greenhouse Gas Emissions Reduction: Solar system only: 1087202 KgCO2; Wind system only: 2459872.5 KgCO2

DALY: Solar system (only): Reduction of 7.1 years; Wind system (only): Reduction of 16.2 years

Species Damage Reduction: Solar system only: 20.4 years; Wind system only: 46 years

Scarcity Damages: Solar system only: $7,266,698; Wind system only: $15,988,564

Reference Part II

Aulisi, A., FARRELL, A. E., PERSHING, J., & VANDEVEER, S. (2005). *GREENHOUSE GAS EMISSIONS TRADING*. Retrieved from World Resource Institute: http://pdf.wri.org/nox_ghg.pdf

Barbose, G., Darghouth, N., Hoen, B., Mills, A., Rand, J., Millstein, D., . . . Oteri, F. (2017). *2017 Wind Technology Markets*. Retrieved from https://www.energy.gov/sites/prod/files/2018/08/f54/2017_wind_technologies_market_report_8.15.18.v2.pdf

Both, A., Manning, T., & Rabin, J. (2014). *Sustainable Farming on the Urban Fringer*. Retrieved from Rutgers University: https://sustainable-farming.rutgers.edu/wp-content/uploads/2014/05/Sample-Electric-Bill-for-Commercial-Customer.pdf

DHS. (n.d.). Retrieved from Department of Homeland Security: https://hifld-geoplatform.opendata.arcgis.com/datasets/electric-power-transmission-lines/data?geometry=-76.933%2C40.133%2C-72.585%2C40.864

DSIRE. (2019). Retrieved from N.C. Clean Energy Technology Center: https://programs.dsireusa.org/system/program/detail/2511

EA REDA. (2019). Retrieved from USDA REDA: https://www.rd.usda.gov/programs-services/rural-energy-america-program-energy-audit-renewable-energy-development-assistance

EERE. (2018). Retrieved from Energy Efficiency and Renewable Energy: https://www.eere.energy.gov/sled/#/results/elecandgas?city=Middletown&abv=NJ&section=electricity&currentState=New%20Jersey&lat=40.3968012&lng=-74.09159820000002

EIA. (2018). Retrieved from U.S. Energy Information Administraion: https://www.eia.gov/state/rankings/?sid=NJ#/series/226

EIA. (2019). Retrieved from U.S. Energy Information Administration: https://www.eia.gov/state/rankings/?sid=NJ#series/31

Electricity Local. (n.d.). Retrieved from https://www.electricitylocal.com/states/new-jersey/

Marsh, J. (2019). *Energy Sage*. Retrieved from Solar Farms: https://news.energysage.com/solar-farms-start-one/

Meehan, C. (2018). *Community Solar Continues to Grow in New York With RooflessSolar Program*. Retrieved from Solar Reviews: https://www.solarreviews.com/news/community-solar-continues-new-york-rooflesssolar-program-041118/

NJ BPU. (n.d.). Retrieved from New Jersey's Clean Energy Program: http://www.njcleanenergy.com/renewable-energy/home/home

nrel. (2018). Retrieved from U.S. Solar Photovoltaic Benchmarck 2018:
https://www.nrel.gov/docs/fy19osti/72399.pdf

RD USDA. (2018). Retrieved from USDA Rural Development Program:
https://programs.dsireusa.org/system/program/detail/917

Solar Maps. (2016). Retrieved from U.S. Department of Energy: https://www.nrel.gov/gis/solar.html

Technical Brief. (2014, July). Retrieved from Enphase energy Systems:
https://enphase.com/sites/default/files/Enphase_PVWatts_Derate_Guide_ModSolar_06-
2014.pdf

USDA. (2017). Retrieved from National Agriculture Statistics Service:
https://www.nass.usda.gov/Publications/AgCensus/2017/Full_Report/Volume_1,_Chapter_2_U
S_State_Level/st99_2_0001_0001.pdf

USDEP. (n.d.). Retrieved from Small Wind Electrical Systems:
https://www.nrel.gov/docs/fy03osti/29948.pdf

Wind Map. (n.d.). Retrieved from U.S. Department of Energy:
https://www.nrel.gov/gis/images/100m_wind/awstwspd100onoff3-1.jpg

Table 1. County Summary Highlights: 2017 (continued)
[For meaning of abbreviations and symbols, see introductory text.]

Item	Hunterdon	Mercer	Middlesex	Monmouth
Farms .... number	1,604	323	217	838
Land in farms .... acres	101,290	25,230	16,023	
Average size of farm ...... acres	63	78	74	47
Median size of farm ...... acres	17	18	10	
Estimated market value of land and buildings:				
Average per farm ...... dollars	986,211	1,414,874	1,607,661	
Average per acre ...... dollars	15,617	18,114	21,773	20,982
Estimated market value of all machinery and equipment $1,000	106,511	26,956	24,444	66,334
Average per farm ...... dollars	66,403	83,438	112,644	79,157
Farms by size:				
1 to 9 acres	392	91	101	318
10 to 49 acres	819	142	77	397
50 to 179 acres	300	51	21	78
180 to 499 acres	71	26	7	26
500 to 999 acres	14	11	9	12
1,000 acres or more	11	2	2	6
Total cropland .... farms	1,216	259	183	592
acres	65,601	15,790	11,246	23,801
Harvested cropland ...... farms	1,112	234	170	527
acres	57,100	12,724	10,052	20,836
Irrigated land .... farms	156	82	77	199
acres	1,835	1,008	2,001	3,550
Market value of agricultural products sold (see text) $1,000	92,246	24,981	38,359	80,633
Average per farm ...... dollars	57,510	77,341	176,772	96,221
Crops, including nursery and greenhouse crops .. $1,000	78,867	20,015	37,593	67,389
Livestock, poultry, and their products $1,000	13,379	4,967	766	13,244
Farms by value of sales:				
Less than $2,500	709	149	93	361
$2,500 to $4,999	197	20	18	94
$5,000 to $9,999	167	35	18	78
$10,000 to $24,999	207	27	20	84
$25,000 to $49,999	82	31	11	64
$50,000 to $99,999	54	21	8	69
$100,000 or more	98	40	40	88
Government payments (see text) ...... farms	94	26	17	28
$1,000	524	149	92	308
Total income from farm-related sources ...... farms	647	123	86	352
$1,000	11,252	2,888	2,513	10,846
Total farm production expenses $1,000	105,833	26,389	36,754	97,971
Average per farm ...... dollars	65,981	81,699	169,373	
Net cash farm income of the operations ...... farms	1,604	323	217	838
$1,000	-1,812	1,630	4,210	9,746

Figure 8: Farm Profile; USDA

Basic Charges					Notes
Customer Number: 1234567890 1234567890 1 - General Service Secondary - GS1 01F					
Customer Charge				$3.25	1
BGS Energy Charges	2,322	KWH x	0.108803	$252.64	2
BGS Transmission Charges	1,121	KWH x	0.005348	6.00	3
	1,201	KWH x	0.005071	6.09	
Total BGS Transmission Charges				12.09	$12.09
BGS Reconciliation Charge	2,322	KWH x	-0.008682	-20.16	4
Delivery Charges	4.9	KW x	3.160000	15.48	5
	10	KW x	0.000000	0.00	
	1,000	KWH x	0.057306	57.37	6
	1,322	KWH x	0.004956	6.55	
Total Delivery Charges				79.40	$79.40
Non-Utility Generation Charges	1,000	KWH x	0.016960	16.96	7
	1,322	KWH x	0.016960	22.42	
Total Non-Utility Generation Charges				39.38	$39.38
Societal Benefits Charges	2,322	KWH x	0.005707	$13.25	
Transitional Assessment Charge	2,322	KWH x	0.002928	$6.80	
System Control Charge	2,322	KWH x	0.000079	$0.18	
Security Deposit Interest				1.86	
Total Charges				$343.78	8
Meter Number	50996554321				
Present KWH Reading	58,836				
Previous KWH Reading	56,514				
Kilowatt Hours Used	2,322				8
Measured Load in KW	11.9				9
Rated Load in KW/KVA	14.9				

¹Customer charge is the fixed monthly service charge.

²Energy Charges, based on **usage**, are typically the largest portion of the bill. This is the fee paid to the generator of the electricity for producing energy (BGS is "Basic Generation Service").

³The two different transmission charges reflect a rate change in the middle of the billing period, in this case due to a difference in rates from summer to winter.

⁴This charge is an adjustment to compensate for the difference between what customers paid for basic generation services and what the utility actually paid the suppliers during the previous month.

⁵The first two delivery charges are the **demand** charges. The method of determining demand can be quite complicated. To understand exactly how a utility calculates **demand**, read the utilities service classification description or contact the utility. Whenever the cost for demand charges is a significant portion of the total bill it is an indication that there may be large equipment that operates for relatively short periods of time. An example might be an irrigation pump. When demand charges are large, consider contacting the utility to explore alternative rate structures. Also, examine the opportunities for reducing peak demands by using smaller equipment or shifting equipment operation to off-peak hours.

⁶In this case, the utility has a relatively large charge for the first 1,000 kilowatt-hours of electricity delivered during the billing period to cover some of the fixed cost of delivering electricity.

⁷The non-utility generation charges and many of the other charges may be described in the notes and definitions included with the electric bill. They are also described in the utilities published rate tariffs. Miscellaneous charges are often small, typically mandated by state government and may be hard to understand. Most utilities will provide a fact sheet that explains typical charges and rates.

⁸Dividing total charges by Kilowatt Hours used gives the cost per unit (in this case 16.5¢ per kilowatt-hour).

Figure 9; Sample electric bill; Rutgers.edu

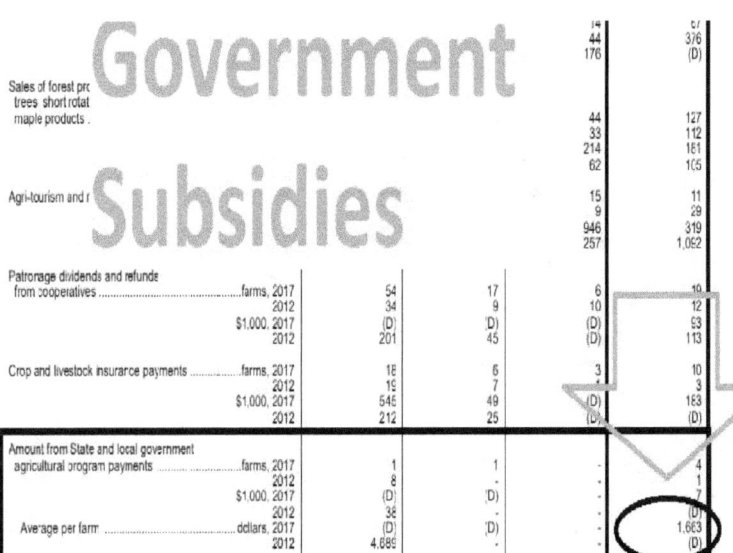

		14	67	
		44	376	
		176	(D)	
Sales of forest products, Christmas trees, short rotation woody crops, and maple products		44	127	
		33	112	
		214	181	
		62	105	
Agri-tourism and recreational services		15	11	
		9	29	
		946	319	
		257	1,052	
Patronage dividends and refunds from cooperatives ..... farms, 2017	54	17	6	10
2012	34	9	10	12
$1,000, 2017	(D)	(D)	(D)	53
2012	201	45	(D)	113
Crop and livestock insurance payments ..... farms, 2017	18	6	3	10
2012	19	7	-	3
$1,000, 2017	545	49	(D)	183
2012	212	25	(D)	(D)
Amount from State and local government agricultural program payments ..... farms, 2017	1	1	-	4
2012	8	-	-	1
$1,000, 2017	(D)	(D)	-	(D)
2012	38	-	-	(D)
Average per farm ..... dollars, 2017	(D)	(D)	-	1,663
2012	4,689	-	-	(D)

Figure 10; Government subsidies for farms; USDA

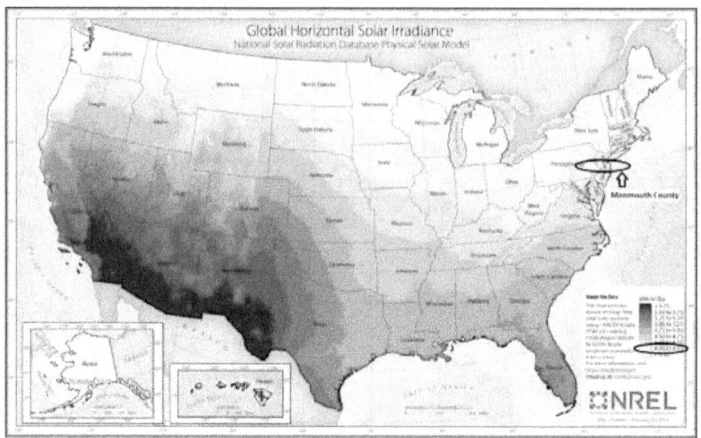

Figure 11: Solar Map, US Dept. of Energy

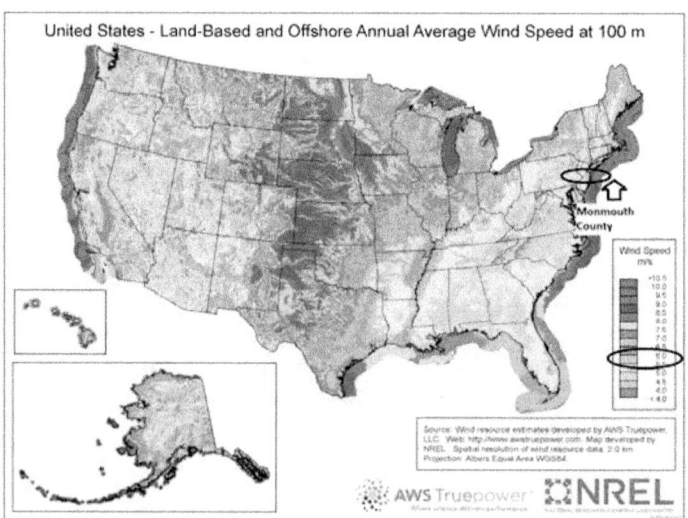

Figure 12: Wind map; US dept. of energy

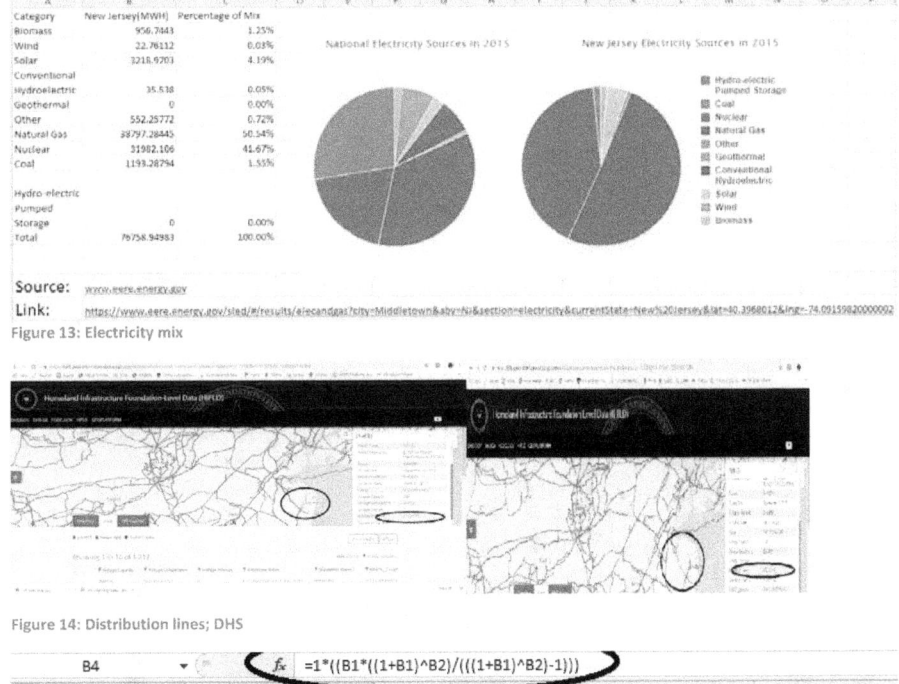

Category	New Jersey(MWH)	Percentage of Mix
Biomass	956.7443	1.25%
Wind	22.76112	0.03%
Solar	3218.9703	4.19%
Conventional Hydroelectric	35.538	0.05%
Geothermal	0	0.00%
Other	552.25772	0.72%
Natural Gas	38797.28445	50.54%
Nuclear	31982.106	41.67%
Coal	1193.28794	1.55%
Hydro-electric Pumped Storage	0	0.00%
Total	76758.94983	100.00%

National Electricity Sources in 2015

New Jersey Electricity Sources in 2015

Legend:
- Hydro-electric Pumped Storage
- Coal
- Nuclear
- Natural Gas
- Other
- Geothermal
- Conventional Hydroelectric
- Solar
- Wind
- Biomass

Source: www.eere.energy.gov

Link: https://www.eere.energy.gov/sled/#/results/electandgas?city=Middletown&abv=NJ&section=electricity&currentState=New%20Jersey&lat=40.3968012&lng=-74.091598200000002

Figure 13: Electricity mix

Figure 14: Distribution lines; DHS

B4   fx =1*((B1*((1+B1)^B2)/(((1+B1)^B2)-1)))

	A	B	C	D	E	F
1	Annual Percentage Rate (annual interest) =	2%				
2	Grace Period (years) =	30				
3						
4	Annual Capital Costs =	0.0446				
5						
6	Annual Operating costs=	$ 78,692				

Rate for a 2% 30 year mortgage ratio

Figure 15; 2% mortgage rate

Size of the array in KW =		2600			
Derate Factor		0.861			

Month	Solar Irradiation in Hrs/day equivalent with 1000 W/m2 (kWh/m2/day)	Days per month	DC Capacity in kW	DC to AC Derate Factor 2018	Average Monthly kWh produced (AC)
January	3.28	31	2600	0.861	227621
February	4	28	2600	0.861	250723
March	4.71	31	2600	0.861	326858
April	4.93	30	2600	0.861	331089
May	5.09	31	2600	0.861	353229
June	5.27	30	2600	0.861	353923
July	5.2	31	2600	0.861	360862
August	5.14	31	2600	0.861	356699
September	4.91	30	2600	0.861	329746
October	4.45	31	2600	0.861	308815
November	3.36	30	2600	0.861	225651
December	3.05	31	2600	0.861	211660

Total Annual Prod = 3636874 KWh/year

Solar Irradiance figures

Select Country	United States
Select State	New Jersey
Select Town/City	Long Branch
Solar Panel direction	Facing directly South

Long Branch Average Solar Insolation figures

Measured in kWh/m²/day onto a solar panel set at a 50° angle (For best year-round performance)

Jan	Feb	Mar	Apr	May	Jun
3.28	4.00	4.71	4.93	5.09	5.27

Jul	Aug	Sep	Oct	Nov	Dec
5.20	5.14	4.91	4.45	3.36	3.05

Efficiency for PV panels in W/m2 =	150

Area for PV in m2 that you need considering 150 W/m2 par 20392.16

Price per Wdc in USD	$	1.78
Cost of the Project	$	7,869,240

Operating costs as a percentage of capital costs	1.00%

Figure 16: Solar worksheet

Enphase Adjustments To PVWatts Default Derate Factors

Derate Category	PV Watts Default (String)	PV Watts Adjusted (Enphase Systems)	PV Design Tools (Enphase Systems)
PV Module Nameplate DC Rating	0.950	0.950	0.950
Inverter and Transformer*	0.920	0.965	0.965
Mismatch*	0.980	0.995 (Max Value)	1.000
Diodes and Connections	0.995	0.995	0.995
DC Wiring*	0.980	0.990 (Max Value)	0.995
AC Wiring	0.990	0.980	0.980
Soiling	0.950	0.970	0.970
System Availability*	0.980	0.995 (Max Value)	0.998
Shading (See Table)	1.000	1.000	1.000
Sun-tracking	1.000	1.000	1.000
Age	1.000	1.000	1.000
Total Derate Factor (Project Efficiency)	0.770	0.850	0.861

Figure 17: Derate factor; PVwatts

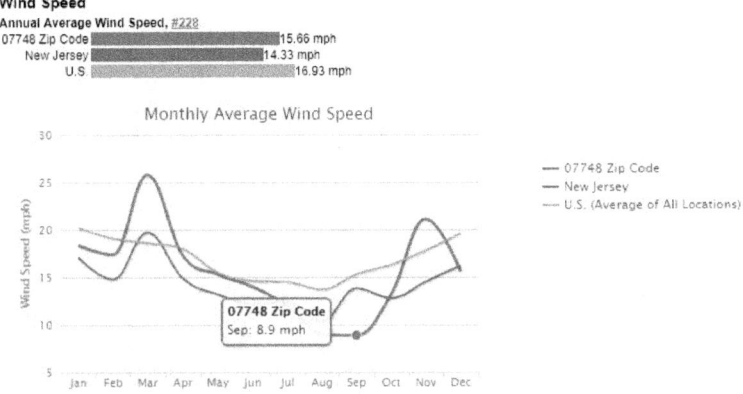

Figure 18: Average wind speed; NREL

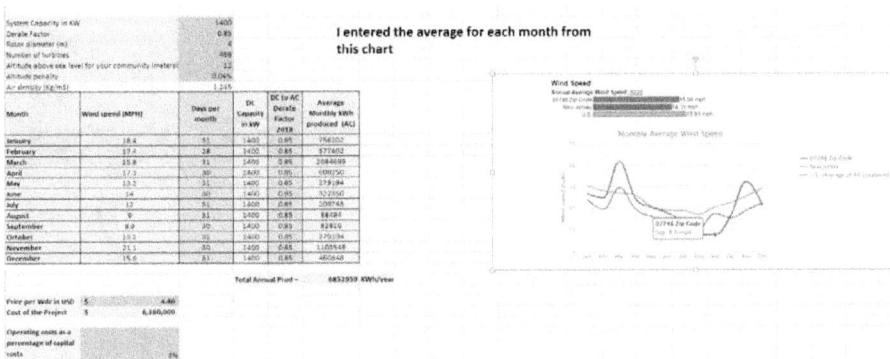

Figure 19: Wind calculator

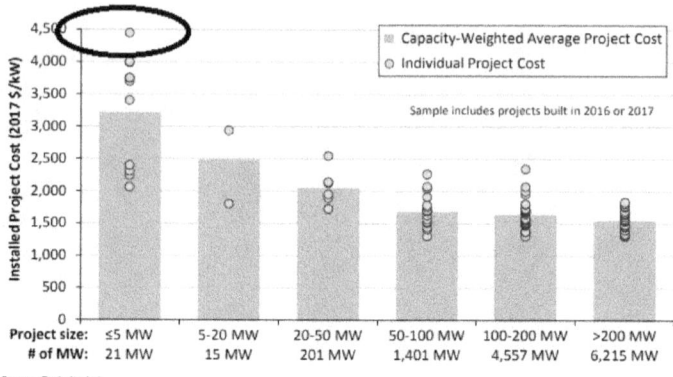

2017 Wind Technologies Market Report
Installed Project Cost (2017 $/kW)
4,500
4,000
3,500
3,000
2,500
2,000
1,500
1,000
500
0
Capacity-Weighted Average Project Cost
Individual Project Cost
Sample includes projects built in 2016 or 2017
Project size: ≤5 MW 5-20 MW 20-50 MW 50-100 MW 100-200 MW >200 MW
# of MW: 21 MW 15 MW 201 MW 1,401 MW 4,557 MW 6,215 MW
Source: Berkeley Lab
Figure 44. Installed wind power project costs by project size: 2016–2017 projects

Figure 20: Wind installaiton cost; Barbos et al

Customer type	System size	Electricity rate schedule
Level 1	Limited to 10kW	9.4 cents/kWh
Level 2	Limited to 2000kW	9.4 cents/kWh
Level 3	Does not qualify under Level 1 or 2 interconnection	9.4 cents/kWh

Figure 21: Electricity purchase price; JCP&L

| 1332 MIDDLETOWN | Street: | | | | | | Year Built: | 2013 |
| 2 | City State: | | | | | | Style: | 5 |

			Additional Information					
	Addl Lots:						EPL Code:	0 0 0
	Land Desc:		3.0 ACRES				Statute:	
0	Bldg Desc:						Initial:	000000 Further: 000000
	Class4Cd:		0				Desc:	
238	Acreage:		3				Taxes:	31202.01 / 31267.33

| | | | Sale Information | | | | | |
| 8761 Page: 4103 | | Price: | 650000 NU#: 0 | | | | | |

Book		Page		Price		NU#	Ratio	Grantee

			TAX-LIST-HISTORY					
Land/Imp/Tot	Exemption	Assessed	Property Class					
330000	0	1441200	2					
1111200								
1441200								
340000	0	1447500	2					
1107500								
1447500								
340000	0	1425400	2					
1085400								
1425400								
340000	0	1409500	2					

Price per acre:
$330,000 land value / $1,441,200 total value = 0.23 (23%)
0.23 * total tax bill $31,202 = $7,176 for 3 acres of land
$7,176 / 3 acres = $2,392 per acre per year

Figure 22: Tax bill; New Jersey Tax Records

Product/Process	Human Health in Disability Adjusted Life Years (DALYs)	Ecosystem Damages in Species.Years (Sp.yr)	Resource Depletion costs in 2008 US Dollars	Carbon Emissions (Metric Tons of CO2eq)
Local Electricity Mix	9.1	25.3	$8,690,927	1346.3
US National Electricity Mix	10.4	42.8	$12,458,852	2276.8
Your phovoltaic production	1.4	3.2	$843,341	172.2
PV reductions compared to Local Electricity Mix	7.7	22.1	$7,847,586	1174.1
PV Reductions compared to U.S. National Electricity Mix	9.0	39.6	$11,615,511	2104.5

Figure 23: CO2 reduction; solar calculator ENVR S-102

Your local Electricity mix is:		1.55%	50.54%	41.67%	0.00%	1.26%	0.06%	4.19%	0.03%	0.00%	0.00%		
		8.016731	0.3455664	0.0053322	0	0.0005721	1.0511E-06	0.001984167	3.372E-06	0		0.3701901	
Impact category	Unit	Coal Electricity 1 KWh	Natural Gas Electricity 1 KWh	Nuclear Electricity 1 KWh	Oil Electricity 1 KWh	Biomass Electricity 1 KWh	Hydro Electricity 1 KWh	Photovoltaic Electricity 1 KWh	Wind Electricity 1 KWh	Geothermal Electricity 1 KWh	Wood Electricity 1 KWh	Your local electricity mix	U.S. Electricity Mix
Climate change	kg CO2 eq	1.0794166	0.6837464	0.0127963	0.9349522	0.045764	0.00370217	0.04735481	0.011239	0.060463722	0.0324561	0.3701901	0.626019272
Ozone depletion	kg CFC-11 eq	8.207E-12	4.261E-10	6.84E-08	3.69E-11	9.498E-16	2.3408E-10	9.41106E-09	7.2E-10	3.36189E-08	3.189E-09	2.911E-09	1.31963E-08
Human toxicity	kg 1,4-DB eq	0.0475572	0.1244781	0.0467851	0.5066356	0.0012946	0.00116454	0.06363280	0.0112764	0.000407195	0.039787	0.0857967	0.070772339
Photochemical oxidant formation	kg NMVOC	0.003701	0.0011924	5.745E-05	0.0027343	0.0402642	1.7091E-05	0.000172941	3.584E-05	1.8616E-05	0.0001722	0.0011895	0.002421115
Particulate matter formation	kg PM10 eq	0.0021061	0.001263	7.629E-06	0.0009581	0.0002207	2.0672E-06	7.93298E-06	3.218E-06	4.25387E-06	8.188E-06	0.0007038	0.001197116
Ionising radiation	kg U235 eq	0	0.0031803	1.0726282	0	0	0.0009892	0.014316984	0.0019678	0.026793117	0.0026868	0.4491722	0.20482476
Terrestrial acidification	kg SO2 eq	0.0086197	0.0060031	8.195E-06	0.0034363	0.0007646	1.3263E-05	0.000296003	4.784E-05	0.000563441	0.0026635	0.0032199	0.00907357
Freshwater eutrophication	kg P eq	0	6.899E-06	5.761E-06	0	0	1.0345E-06	3.9012E-06	7.225E-06	0.000225642	4.719E-06	7.047E-06	3.1791E-06
Marine eutrophication	kg N eq	0.0001201	2.517E-05	6.402E-06	8.608E-05	3.589E-06	7.387E-07	1.81724E-06	4.219E-06	3.96196E-06	3.679E-06	1.846E-06	6.729018E-06
Terrestrial ecotoxicity	kg 1,4-DB eq	1.569E-05	0.0001227	6.823E-06	1.812E-05	8.038E-09	7.2417E-07	9.36272E-06	1.309E-06	0.00135156	0.0002488	6.851E-05	4.00269E-05
Freshwater ecotoxicity	kg 1,4-DB eq	0.0001175	0.0086372	0.0004263	0.0032378	8.311E-06	4.0659E-05	0.00003977	0.0003579	0.001227056	0.0001663	0.0045796	0.00263860
Marine ecotoxicity	kg 1,4-DB eq	0.0001487	0.0025932	0.0006433	0.0036733	8.57E-06	4.199E-05	0.001029106	0.000377	0.00502953	0.000204	0.0015372	0.000921843
Agricultural land occupation	m2a	0	0.0001527	0.0004878	0	0	4.4683E-06	0.001876087	0.0002229	0.001000769	2.089267	0.0003691	0.001007151
Urban land occupation	m2a	0	0.00014	0.0002766	0	0	4.2695E-05	0.000356238	0.0008741	0.000725619	0.003727	0.0002012	0.00014553
Natural land transformation	m2	0	3.328E-06	2.326E-06	0	0	-1.061E-07	0.98929E-06	1.286E-06	1.75121E-05	3.1936E-06	3.298E-06	1.63078E-06
Water depletion	m3	0	4.199E-06	0.003263	0	0	4.627E-06	0.000489819	0.0001114	0.013023488	0.0010642	0.0013973	0.00084269
Metal depletion	kg Fe eq	0	0.0005689	0.0065967	0	0	0.00186796	0.016654131	0.0140824	0.017421036	0.0016391	0.0037931	0.00218901
Fossil depletion	kg oil eq	0.3362224	0.2795956	0.0036057	0.3170306	0.000816	0.00008845	0.014364748	0.0034682	0.016721311	0.0087308	0.1480869	0.213164861

ELECTRICITY MIX

Figure 24: Electricity mix char; solar calculator ENVR S-102

Product/Process	Human Health in Disability Adjusted Life Years (DALYs)	Ecosystem Damages in Species.Years (Sp.yr)	Resource Depletion costs in 2008 US Dollars	Carbon Emissions (Metric Tons of CO2eq)
Local Electricity Mix	9.1	25.3	$8,690,927	1346.3
US National Electricity Mix	10.4	42.8	$12,458,852	2276.8
Your phovoltaic production	1.4	3.2	$843,341	172.2
PV reductions compared to Local Electricity Mix	7.7	22.1	$7,847,586	1174.1
PV Reductions compared to U.S. National Electricity Mix	9.0	39.6	$11,615,511	2104.5

Figure 25: Health & Environment benefits; solar calculator ENVR S-102

Product/Process	Human Health in Disability Adjusted Life Years (DALYs)	Ecosystem Damages in Species.Years (Sp.yr)	Resource Depletion costs in 2008 US Dollars	Carbon Emissions (Metric Tons of CO2eq)
Local Electricity Mix	17.065	47.694	$16,376,307	2536.9
US National Electricity Mix	19.577	80.654	$23,476,205	4290.1
Your wind production	0.908	1.448	$387,743	77.0
Wind reductions compared to Local Electricity Mix	16.156	46.246	$15,988,564	2459.9
Wind Reductions compared to U.S. National Electricity Mix	18.669	79.206	$23,088,462	4213.1

Figure 26: Health & Environmental benefits; wind calculator ENVR S-102

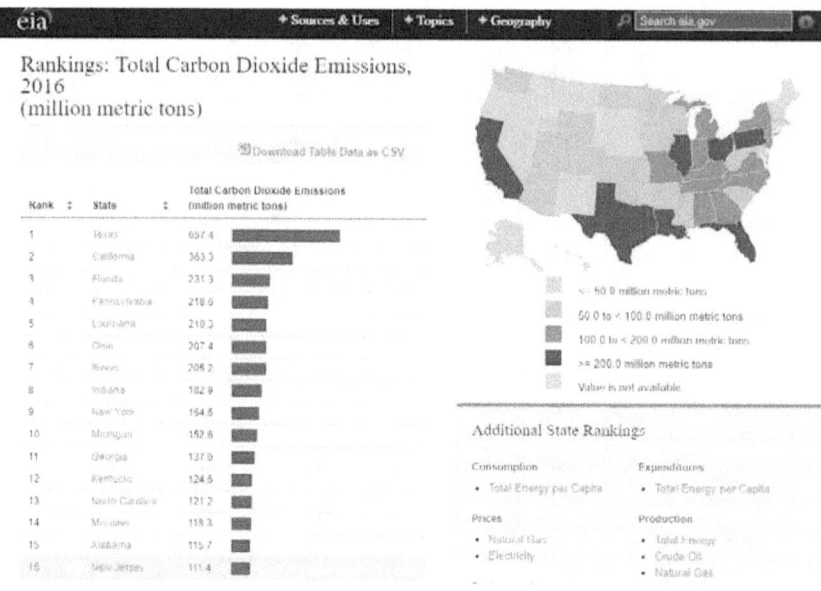

Figure 27: Rank of NJ CO2 emission; EIA

Publisher: Eliva Press SRL

Email: info@elivapress.com

Eliva Press is an independent publishing house established for the publication and dissemination of academic works all over the world. Company provides high quality and professional service for all of our authors.

Our Services:
Free of charge, open-minded, eco-friendly, innovational.

-Free standard publishing services (manuscript review, step-by-step book preparation, publication, distribution, and marketing).
-No financial risk. The author is not obliged to pay any hidden fees for publication.
-Editors. Dedicated editors will assist step by step through the projects.
-Money paid to the author for every book sold. Up to 50% royalties guaranteed.
-ISBN (International Standard Book Number). We assign a unique ISBN to every Eliva Press book.
-Digital archive storage. Books will be available online for a long time. We don't need to have a stock of our titles. No unsold copies. Eliva Press uses environment friendly print on demand technology that limits the needs of publishing business. We care about environment and share these principles with our customers.
-Cover design. Cover art is designed by a professional designer.
-Worldwide distribution. We continue expanding our distribution channels to make sure that all readers have access to our books.

www.elivapress.com

www.ingramcontent.com/pod-product-compliance
Lightning Source LLC
Chambersburg PA
CBHW051255170526
45165CB00004B/1724